Anastasia Shchurenko

Building an ecosystem and infrastructure for smart confectionery
Book 1

Anastasia Shchurenko

Building an ecosystem and infrastructure for smart confectionery Book 1

Innovative transformation of the ecosystem and infrastructure of a smart composite manufacturing facility

ScienciaScripts

Imprint

Cover image: www.ingimage.com

This book is a translation from the original published under ISBN 978-620-7-47294-9.

Publisher:
Sciencia Scripts
is a trademark of
Dodo Books Indian Ocean Ltd. and OmniScriptum S.R.L publishing group

120 High Road, East Finchley, London, N2 9ED, United Kingdom
Str. Armeneasca 28/1, office 1, Chisinau MD-2012, Republic of Moldova, Europe
Printed at: see last page
ISBN: 978-620-7-87394-4

Anastasia Shchurenko

Book - 1

Caption

Building a smart confectionery production ecosystem. Book 1

Subheading

Innovative transformation of the ecosystem and infrastructure of a smart manufacturing facility for composite products.

Keywords:

Confectionery production; Confectionery production ecosystem; Smart confectionery production; Smart confectionery production infrastructure and its modules; Hydrodynamic in-flow mixing; Mixing with simultaneous kneading; Homogenisation of liquid confectionery components; Dynamic impact process on mixed liquid confectionery components; Active, periodic kinetic regime.

Annotation

Innovative transformation of the ecosystem and infrastructure characteristics of a smart manufacturing facility into an advanced combined supersystem with interconnected subsystems of on-line real-time control of automatic energy and water treatment modules. The supersystem includes preliminary analysis of long-term forecasts of development of this field of technology, use of the results of the analysis to form the process of integration of 44 commercial system criteria for evaluation of a smart object of confectionery production with systems of 40 + 10 criteria and methods of obtaining an ideal final result according to the qualification of the Theory and Algorithm of Inventive Problem Solving.

Specificity of design and logistics of construction of the ecosystem of the object or smart complex of confectionery production and its innovative components of the ecosystem as an autonomous supersystem and incoming autonomous innovative infrastructure subsystems, ensuring its maximum economic efficiency and environmental friendliness of the processes of construction, adjustment and operation, with maximum saving of energy and other resources and full control of all basic parameters, built on electromagnetic resonance spectroscopy.

Table of Contents

Introduction

Confectionery products have traditionally been the most important basic part of the food programmes of various consumer cultures and, therefore, the development of this group of products and technologies for their production and consumption is currently receiving the greatest attention;

The emergence and detailed examination of the complex problems associated with so-called smart confectionery products show that the use of sugary foods requires in turn increased attention, including special production technologies and the provision of safe consumption programmes

Confectionery products are closely related to the production and consumption of dairy products and, as a rule, the technology of dairy products and the technology of production of all kinds of variants and versions of confectionery products have a very significant mutual influence and very often innovations in dairy production initiate the development of innovative technologies and products in the confectionery sector of food production;

In this regard, special importance is attached to the application of elements of artificial intelligence and artificial neural networks.

The importance attached to this issue is evidenced by the fact that, for example, in the European Union, legislation is being drafted to control and manage the implementation of these technologies;

The European Union has agreed on the world's first draft law on the regulation of artificial intelligence (AI). After three days of negotiations, representatives of the European Parliament, the European Commission and EU member states decided on the categories of risks that the use of AI can create: minimal, limited, high and unacceptable.

According to the adopted document, the use of "unacceptable risk" systems will be prohibited.

These included systems such as:

— affecting the human subconscious,

— used to identify people's vulnerabilities, including by age, disability, social or economic status,

— Emotion recognition systems by employers and educational institutions,

— Systems that produce social ratings based on a person's social behaviour or personal characteristics.

At the same time, the negotiators agreed on a number of exceptions for the use of biometric identification systems in public places. In particular, the use of such systems by law enforcement agencies is allowed for a certain list of offences and only with the permission of the court.

The remaining categories will be subject to different requirements depending on the level of risk.

A separate section is dedicated to respecting consumer rights: it will be possible to file complaints about incorrect operation of the AI. There are fines for violations.

"This is a historic achievement and a huge milestone on the road to the future! In this endeavour, we have managed to maintain an extremely delicate balance: to foster innovation and the adoption of artificial intelligence across Europe, while fully respecting the fundamental rights of our citizens", is the opinion of the Spanish Secretary of State for Digitalisation and Artificial Intelligence ;

According to the EU commissioner," it is a launching pad for EU startups and researchers to lead the global artificial intelligence race.

For the so-called smart modern production of confectionery products, as an example, a fundamentally new technology of hydrodynamic mixing of different liquid components with simultaneous knocking down in a dynamic, periodically repeating, active kinetic mode has been developed

Figure 1

The figure shows a photo of the experimental module for mixing and simultaneous knocking down of liquid products;

The proposed technology is a process of dynamic action on the liquid or partially consistent components to be mixed and knocked down;

Figure 2

The figure shows an experimental and test version of a manually driven module for homogeneous mixing and preparation for fermentation processes of dairy product components, e.g. butter.

The most researched in this process are components from various dairy products , which can also be used in smart confectionery and its infrastructural elements ;

The aim of this research is to find new ways for the production of innovative composite food products, which have no or minimised fat content and which, on the contrary, have increased concentrations of vitamins and biologically active elements;

An additional purpose of the proposed process is the possibility, without heat treatment and without other types of influence on the product, as already mentioned, mainly dairy or lactic acid, to obtain a mixture or whipped mixture of components, which under normal conditions are either not mixed at all, or poorly mixed;

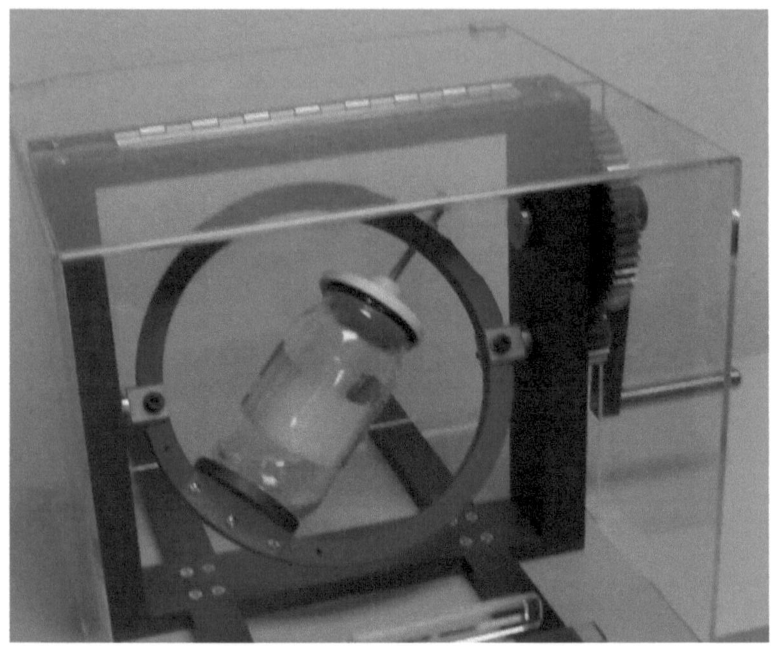

Figure 3

The figure shows a photo of the experimental module

In order to be able to apply the proposed technology in the conditions of modern production, the next goal is to be able to obtain a stable and stable mixture that maintains its properties and homogeneity for a long time;

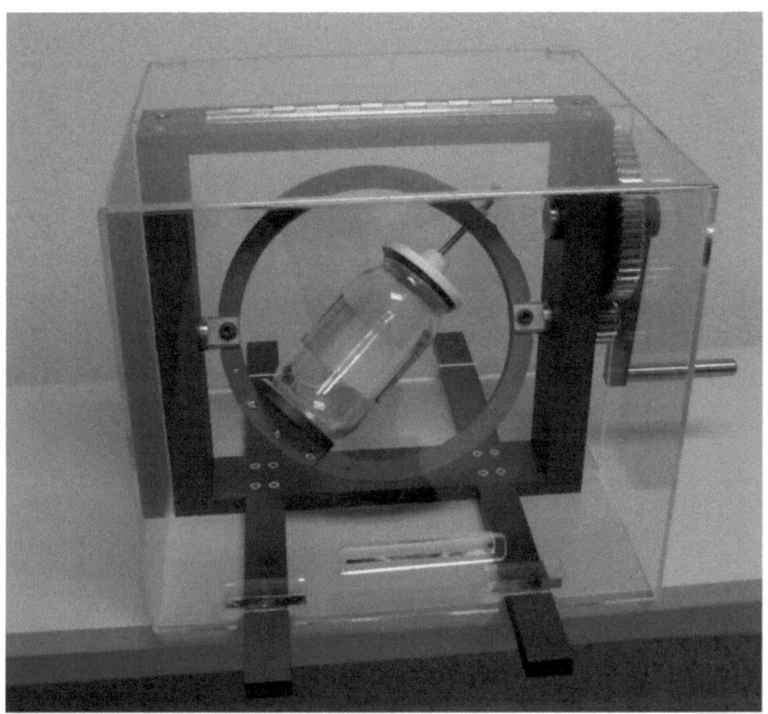

Figure 4

The figure shows an experimental version of the module for testing the conceptual solution; an additional difference in its design is the possibility to change the angle of inclination of the tank in which the components to be mixed and knocked down are placed; in addition, the module has the possibility to change the gear ratio of the drive to rotate the tank with the processed components of the mixture.

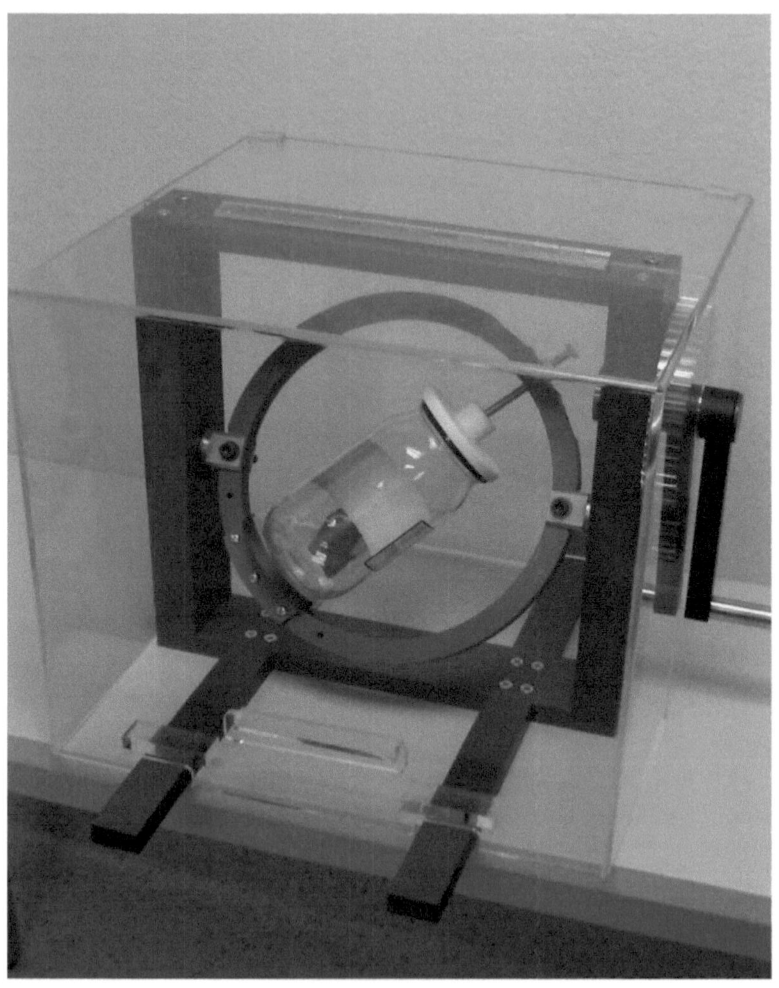

Figure 5
One of the design variants of the experimental module

Figure 6

One of the design variants of the experimental module

The end products of the proposed technology are:

- butter, made from sour cream, with an unusually low fat content of no more than 13%;
- butter, made from sour cream with multivitamin additives, - a mixture of sour cream with carrot juice;
- butter beaten with sour cream and tomato juice;
- butter beaten with sour cream and various fruit syrups;
- butter beaten with sour cream and honey;
- Butter beaten sour cream with garlic, dill;
- Creamy butter made from sour cream modified with egg butter;
- butter beaten from sour cream modified with alpha-lecithin;
- butter beaten with sour cream modified with albumin;

- butter beaten with sour cream modified with lysozyme;
- butter, made from sour cream, modified with cocoa butter;
- Creamy butter beaten with sour cream, modified with avocado paste;
- Butter beaten with sour cream modified with vegetable pastes;
- butter beaten with sour cream modified with fruit pastes;

Direct processing of milk and its modification when mixed with other components, allows to create new modified combined milk composites with new, unusual properties and exceptionally high consumer quality.

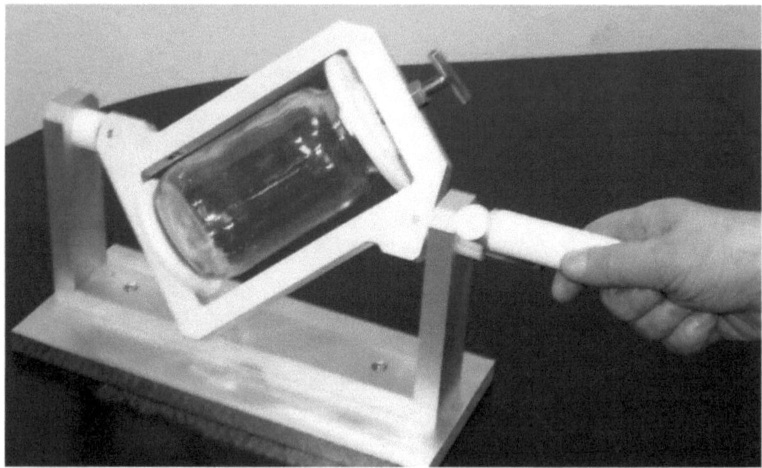

Figure 7

First prototype of an experimental module for proof-of-concept testing of the technology

Figures 8 and 9

First prototype of an experimental module for proof-of-concept testing of the technology

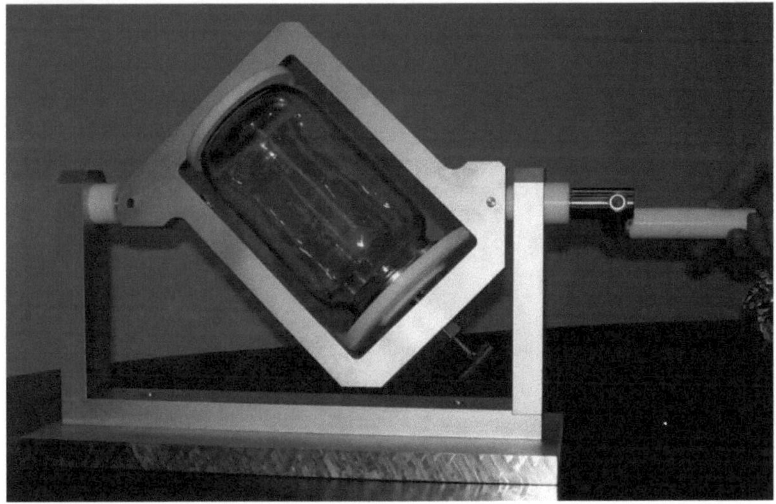

Figure 10

First prototype of an experimental module for proof-of-concept testing of the technology

First of all, it is a whipped mixture of milk and honey; moreover, for the mixture is used skimmed milk - 1% fat content, cooled to a temperature of 2-3 degrees Celsius and natural bee honey in the proportion of - 75% of milk and 25% of honey; the resulting mixture is stable, homogeneous, has an exceptionally fine, pleasant flavour and is very useful;

Thanks to the wide range of possibilities of the hydrodynamic mixing and kneading machine, it is possible to prepare mixtures of many variations and combinations of materials, liquids, pastes and syrups;

The apparatus implementing the new technology can be manufactured for a wide range of outputs. One option, with a capacity of approximately 1 litre or 850 grams of butter per beating cycle, could be an appliance for farmers, which could be used to process surplus milk;

The apparatus is based on an invention which is based on the principle of forming hydrodynamic kinetic linear reciprocating pulsations in a closed volume of a container with the components to be mixed and dislodged. The initiated movements are so effective that they allow to achieve the mixing and dislodging effect in a relatively short period of time, with an exceptionally homogeneous structure of the resulting mixture or oil;

The appliance can be designed as a mass-market product, with a small capacity of approximately 1 litre for the ingredients to be mixed and beaten; it can be electrically driven, but it can also be manually driven, which will significantly reduce its price and increase its sales;

The appliance can be used as an accessory for cafes and restaurants, in which case its capacity can reach up to 2 litres and it should come with several interchangeable containers for greater flexibility and versatility;

The device can be used as Original Process Equipment for industrial enterprises engaged in processing of dairy products and can be used for production of small series of dairy food composites;

The device can be designed as an automated technological equipment for large-scale and mass production of modified dairy products;

Principle differences between the proposed butter churning process and known processes for the same purpose :

- The mixing is carried out in a flask in the shape of a cylinder, closed at the ends by hemispherical lids with conical holes in the axis;
- the knockdown process with a similar level of efficiency may be carried out in equivalent flask embodiments to said embodiment;
- the whole structure consisting of a cylindrical jacket with conical sharpenings on its ends and two hemispherical lids having conical chamfers on the base and coaxial to them conical holes on the axis, assembled and sealed by inserting into the conical holes of the lids, previously aligned with the cylindrical jacket by conical sharpenings on its ends and conical chamfers on the base of the lids, - clamping, orienting, fixing and sealing the internal volume of the said structure, fixing clamping devices installed in the lids.
- Thanks to these design features, in addition to the significant differences, the churning process also benefits from significant advantages in terms of the quality of the

product obtained and the possibilities of the technological range of the oil churning process;

- additional possibilities in the process realised by the proposed device:

- The volume in which the oil is churned is collapsible;

- in the process of knocking down there is a possibility of authorised penetration into the internal volume, introduction of the components of the mixture to be knocked down, removal of waste or part of the components from the working volume;

- conical holes of hemispherical lids contribute to a more complete breaking of the mixture when hitting the inner hemisphere of the lids, i.e. they fulfil a number of combined functions, such as activating, combining and orienting, sealing, fixing, and, most importantly, they fulfil the functions of a technological gateway on both sides of the structure, both outlet and inlet, depending on the position of the flask;

- the specified design allows to have several sets of flasks on one device for mixing, which allows, at a relatively short duration of the mixing process, to change the composition of components for mixing, including the nature of lactic acid matrix, the degree of its fat content and other parameters;

- hydrodynamic parameters of the process of knocking down, in the form of flask elements, in the presence at the end of each linear movement of the knocked down mass of a spherical recess with an activating cone in the centre, allow to significantly increase the efficiency of the process of knocking down, which is explained by the fact that the volume of the knocked down mass at each impact in the lid, occurring after half a turn of the leash frame, is transformed, passing within a very short period of time from the cylindrical form to the spherical, that is, there is a process of changing the form and working volume

- The described character of influence on the mass to be knocked down, due to the fact that there is a kinetic variety of types of forced changes in the volume and geometric shape of the mass to be knocked down, under the influence of many force factors caused by the combination and concentration of movements and loads caused both by the shape of the flask geometry and the character of movements caused by the kinematics of the system of the apparatus for oil knocking, allow, in contrast to known methods, to obtain a homogeneous composite product as a result of the process realisation.

From the presented models of hemispherical lids, it can be seen that :

- their design is technologically advanced;

- they are highly durable;

- they are easy to maintain;

- they're hygienic;

- They can be made from a wide variety of materials;

- for the same cylindrical jacket design it is possible to have different variants of covers, differing in the steepness of the sphere and the depth of the conical hole along the axis, while maintaining the identity of the jointing elements of the design.

- in the distinctive features concerning the apparatus, a consequence is seen, which comes from the fact that the frame can be of various shapes, and that, when it is made, as a variant, in the form of a flat ring with threaded holes made along one of the geometrical coordinates of this ring, from the periphery of the ring to the centre, as can be seen from the model, and, when performed on the plane of the disc at least two holes for fixing the disc in the spindle supports, when the number of holes is more than two, the bulb can be installed at any angle to the axis of rotation of the frame, by changing the holes , which are selected for fixing in the spindle supports;

- the method specifies that the filling or feeding of starting materials into the inner volume of the flask and the removal of liquid waste therefrom are carried out in the assembled state of the flask, through the axial openings of the fixing devices, which for this purpose have threaded plugs closing these openings; when feeding starting materials into the same openings, aerodynamic or any other activating or stimulating devices are introduced, if necessary, which after fulfilling their functions are removed from the inner volume of the flask through the said openings.

- the method specifies that the introduction of additional starting materials can be carried out during the oil knocking process, when the frame rotation is temporarily stopped;

- the method indicates that, if necessary, waste or components that have exhausted their function can be removed from the internal volume of the flask, when the rotation process is interrupted and the flask frame is temporarily stopped;

Thus, in the general distinctive features of the process there are such additional features in addition to the directly stated ones:

- The flask can be set at any angle to the spindle axis; the angle can be varied depending on the viscosity, consistency, composition, temperature, density, acidity level, quantity, volatility, specific gravity, mass proportions of the components of the starting material;

- possibility of input and output of components during the knockdown process;

- the possibility of penetrating the internal volume during the knockdown process;

- possibility of changing or supplementing the composition of ingredients during the beating process, possibility of introducing various flavouring additives, stimulants, emulsifiers, etc. at any time during the beating process;

- the possibility of introducing odour bouquet-forming substances in the finished product, including pheromones or other aroma and flavour-forming ingredients and substances, at any time during the knockdown process;

- possibility of introducing various flavour-forming ingredients, including synthetic ones, allowing to obtain taste sensations unusual for the initial materials introduced into the process;

From the basic process proposed in the invention, as a consequence of the extension of the constructive and technological connections between the main features of the process, additional features for the apparatus also emerge.

Modelling of the element base of the experimental module

Figure 11

The figure shows the elemental base of pipelines in which non-contact control of liquid components parameters is carried out during their transport to the mixing and mixing module or to the fermentation preparation module

The control is based on the principles of electromagnetic resonance spectroscopy at the section located in the centre of symmetry of the pipeline;

The control is carried out by several variants of the resonant non-contact control technology, including a variant with a solenoid sensor and a sensor in the form of a multilayer printed circuit board, which provides the necessary level of signal penetration into the flow of the measured liquid, with the highest possible speed of signal penetration and reflection (usually the duration of the complete measurement cycle is 10 milliseconds).

The printed circuit board - sensor is made by RITM technology (dimensional, selective metal etching) and its thickness is only 50 microns.

Figure 12

The figure also shows the elemental basis of pipelines in which non-contact control of liquid components parameters is carried out during their transport to the mixing and mixing module or to the fermentation preparation module.

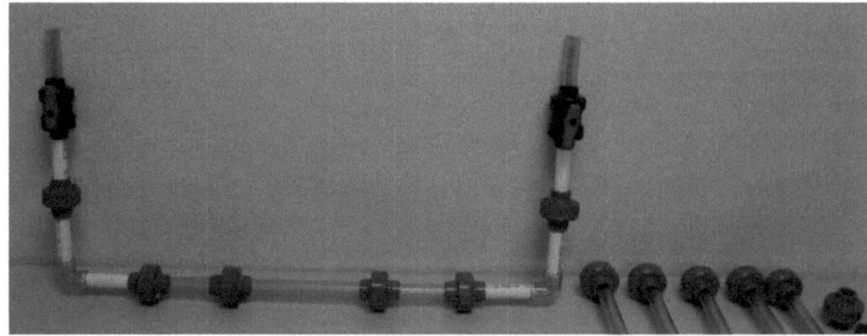

Figure 13

The figure also shows the elemental basis of pipelines in which non-contact control of liquid components parameters is carried out during their transport to the mixing and mixing module or to the fermentation preparation module.

Figures 14 and 15

The figures also show the elemental basis of pipelines in which non-contact control of liquid components parameters is carried out during their transport to the mixing and mixing module or to the fermentation preparation module.

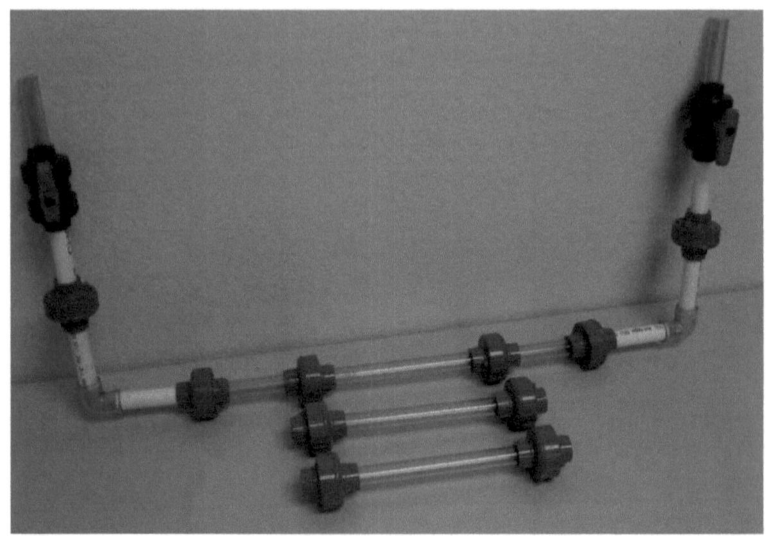

Figure 15 -1.

The figure also shows the elemental basis of pipelines in which non-contact control of liquid components parameters is carried out during their transport to the mixing and mixing module or to the fermentation preparation module.

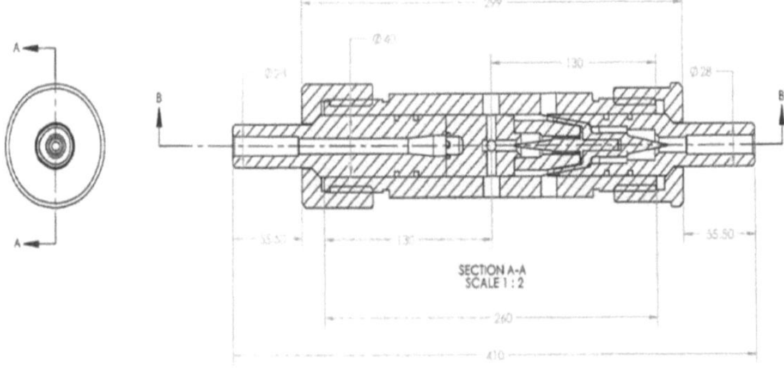

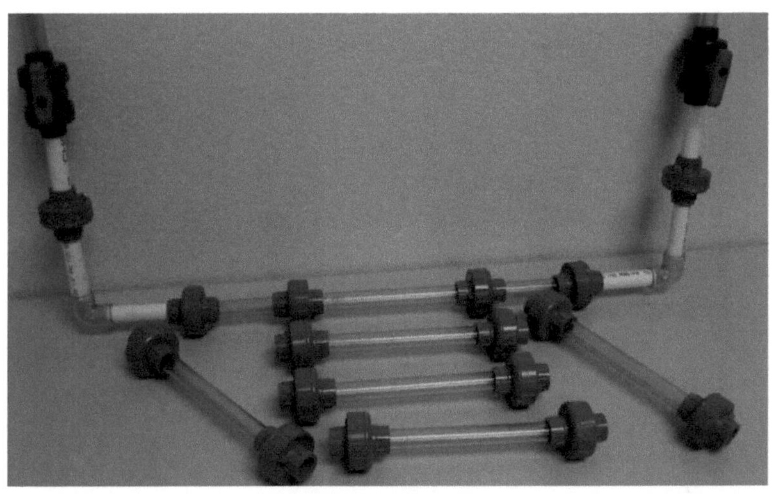

Figures 16 and 17

The figure also shows the elemental basis of pipelines in which non-contact control of liquid components parameters is carried out during their transport to the mixing and mixing module or to the fermentation preparation module.

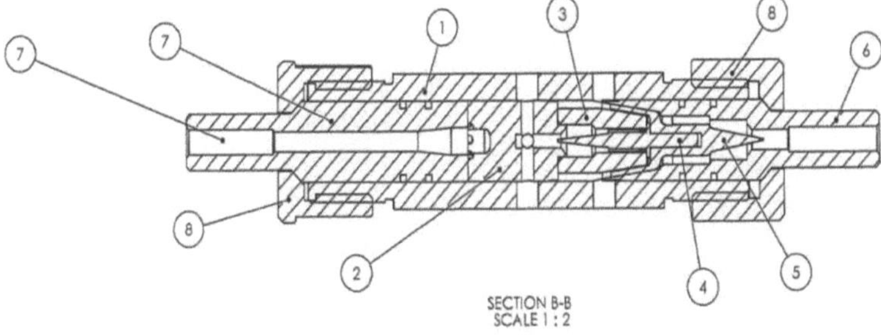

Figure 18

The figure also shows the elemental basis of pipelines in which non-contact control of liquid components parameters is carried out during their transport to the mixing and mixing module or to the fermentation preparation module.

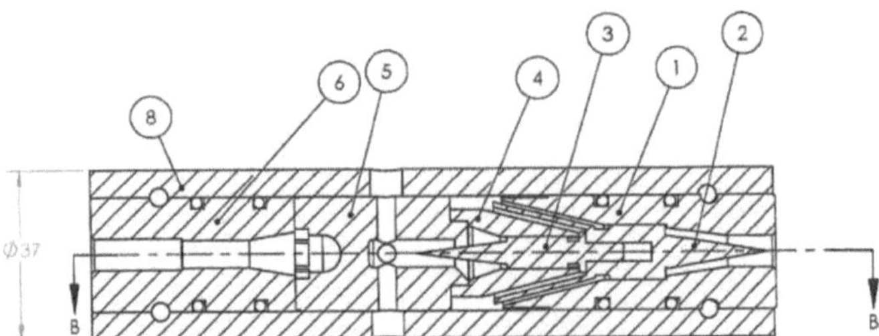

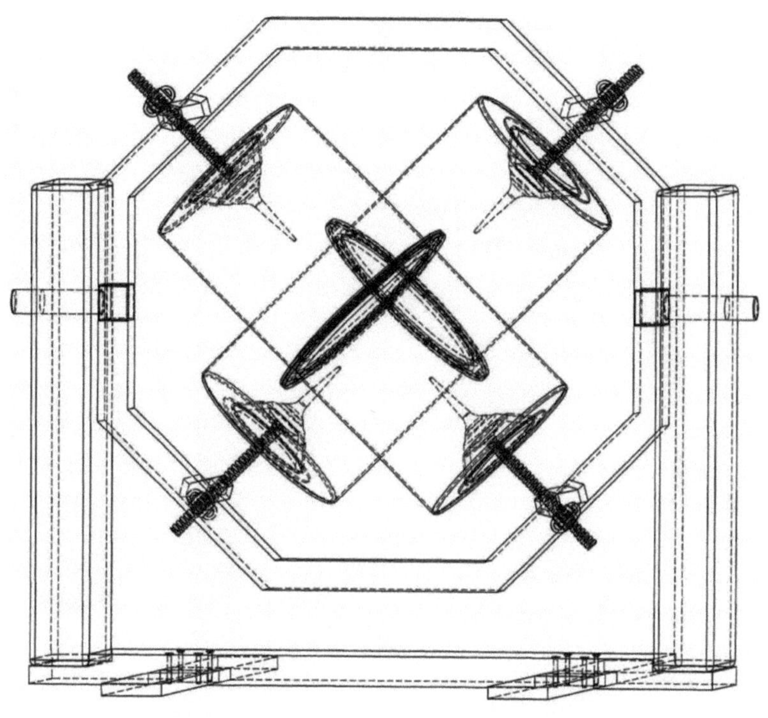

Figure 19

This is a prototype of an adaptable machine for dynamic synthesis of food composites and compositions (the universal working capsule is not shown, it has been dismantled)

Figure 20

It is also a prototype of an adaptable machine for dynamic synthesis of food composites and compositions (the universal working capsule is not shown, it has been dismantled)

Figure 21

Food composite based on naturally fermented lactic acid fat matrix mixed with carrot paste in the proportion of 60 parts of lactic acid matrix per 40 parts of other composite components, on the photo on the left and, on the photo on the right, - food integrated composite based on naturally fermented lactic acid fat matrix dynamically and homogeneously mixed with natural tomato paste and natural garlic paste sauce in the proportion of 70 parts of lactic acid matrix per 30 parts of other composite components;

Figure 22

Food composite on the basis of naturally fermented lactic acid matrix and dynamically mixed with it natural bee honey with addition of nut paste, in the initial proportion of 70 parts of lactic acid matrix to 30 parts of other components of the composite.

Figure 23

Food composite based on a homogeneous mixture of naturally fermented, natural, absolutely free of impurities, without preservatives and stabilisers, lactic acid matrix and a dynamic homogeneous mixture of poppy seeds, previously homogeneously mixed with natural bee honey and natural nut paste, in the initial proportion of 70 parts of lactic acid composition to 30 parts of the other components of the composite.

Apparatus for knocking down food composites including composite oil - description of the apparatus and principles of its real application

The appliance for beating and homogenising food compositions and composite butter is designed for efficient beating of these complex products from lactic acid raw materials, predominantly sour cream or cream, and predominantly low fat content. The starting products or components may be initially subjected to fermentation;

Preliminary tests of the device have shown the possibility of effective beating not only from pure sour cream, but also from sour cream with various additives mainly of vegetable origin.

Such additives . for example, may include the following materials:

- carrot juice;

- carrot puree or paste;

- tomato juice;

- tomato paste;

- cucumber brine;

- cucumber puree;

- mashed avocado;

- mashed kiwi;

- mashed sweet peppers;

- various courgette purees;

- various fruit purees, - apple; plum; peach; pear; apricot; pineapple; Puree of tropical fruits and vegetables;

- various combinations of all the above additives, and in combination with dill, garlic, onions, lemons;

- with prune pulp;

- jam from black rowanberries, sea buckthorn and other medicinal berry plants;

- various purees of strawberries, strawberries, blueberries; blackcurrants; raspberries;

mulberries; kalina; gooseberries and other berries;

The above list can be continued and various vegetables and fruits from local geographical areas can be introduced, also in various combinations and combinations with known products.

It is possible to modify composite oil by introducing vegetable fats in the mixing process in order to change the balance of vegetable and animal fats in the oil;

In order to extend the scope of application of the butter churner, the design of the butter churner allows it to be flexibly adapted to produce butter with a wide variety of additives and in a wide variety of volumes.

Versatility, as the ability to quickly change to knocking down composite oil with other components, is one of the main operational advantages of the apparatus and the knocking down process itself.

The design of the mixer flask , having three constituent elements, of which the original and maximally influencing the results of mixing and obtaining commercial products with different properties are :

- jacket of the mixer flask, having conical recesses on the flanges, for sealing the internal volume and simultaneously for centring and volumetric orientation of the flask geometrical parameters in three-dimensional space, allowing to keep all geometrical spatial relations, including relations of the apparatus axis position, unchanged during the mixing process;

- conditionally upper hemispherical lid having two conical surfaces located on one axis, intended for centring, orientation and sealing of the internal volume of the flask, and allowing to introduce all necessary components of the initial composition for oil knocking into the upper one of them, taking the axial force for fixation and local sealing of the flask from the fixer with an axial overlapping hole, when the flask structure is assembled and sealed.

- conditionally bottom hemispherical lid having three conical surfaces located on the same axis, designed for centring, orientation, sealing and freeing from the waste from the knocking down of the internal volume of the flask, and allowing the waste generated from the oil knocking down process to be discharged into the lower one of them, taking the axial force for fixing and local sealing of the flask, from the fixer with a ball valve, when the flask structure is assembled and sealed.

The process of churning high quality, extra pure, homogenised butter without additives consists of the following operations :

After washing from the previous mixing cycle, the flask and the upper and lower hemispherical lids are pre-matched on the conical surfaces, placed in the bracket and the clamps begin to compress along the axis, until the conical surfaces - two in the upper lid and jacket of the mixer flask and two in the lower lid and jacket of the mixer flask - are completely aligned.

Then open the sealing plug in the upper retainer and pour the required volume of sour cream or fermented cream into the inner volume of the flask.

Then close the stopper in the retainer and begin to rotate the frame with the flask. The sour cream moves along the axis of the inner cavity of the flask, making one complete movement with a bump in the bottom of the lid for every half turn of rotation of the frame.

When the churning cycle is complete, the system is stopped and the frame is rotated and the flask is placed in a position where the bottom lid is at the bottom, the ball valve is opened and the churning waste (by-product) is poured out of the inner cavity of the flask.

After that the clips are released and the flask with caps is removed from the frame; then both caps are removed and the dislodged oil is taken out of the inner cylindrical cavity of the jacket of the mixer flask. The three flask components are then washed and a replacement set is installed in their place.

In case it is necessary to introduce additives or a combination of additives into the oil, all components are introduced according to the above scheme.

If components are to be inserted sequentially or at different intervals, the frame must be stopped and the locking plug opened before each component is inserted.

The possibility of flexible introduction of components into the process, with orientation either by time or by the degree of oil readiness, or by any technological reaction, e.g. by temperature change or by chemical-technological features, considerably extends the technical and technological range of the oil churning apparatus.

PROCESS AND APPARATUS FOR CHURNING ANIMAL COMPOSITE BUTTER FROM MULTI-COMPONENT RAW MATERIALS HAVING COMPONENTS OF PLANT AND ANIMAL ORIGIN EMBEDDED IN LACTIC ACID MATRIX IN DIFFERENT PROPORTIONS

What is invented:

1. A process for churning animal butter, mainly from multi-component raw materials on the basis of a lactic acid matrix with any including reduced fat content,

   Incorporating :

   formation of the starting material in the inner cavity of the mixer flask by sequential introduction of lactic acid matrix and additives of vegetable and animal origin;

   installation of the mixer flask into the mixer leash frame and sealing of its internal volume by means of axial fixation of hemispherical lids;

   mounting of the mixer leash frame in the drive bracket of said mixer, ensuring a certain angle between the axes of the mixer flask and the axis of rotation of the spindles of the drive bracket, and ensuring that this angle remains constant during the entire beating period;

   reciprocating movement of the source material along the axis of the mixer flask with introduction of the mass of the said material into impact contact with the inner surface of the hemispherical lids every half a turn of rotation of the spindles of the drive bracket;

   triggering a change in the direction of flow of the source material for every half turn of the drive clamp spindles;

   Accumulation of kinetic energy of the moving mass of the source material at each change of its direction of motion;

   drainage of process waste from the knockdown process;
   depressurisation of the mixer bulb and its removal from the mixer lead frame;

   freeing the cylindrical jacket of the mixer flask from the hemispherical lids and removing the knocked down product from the cylindrical jacket of the mixer flask.

2. Apparatus for churning animal butter from multi-component raw materials based on lactic acid matrix and additives of vegetable and animal origin,

Consisting of permanent and replaceable elements, whereby permanent elements contain spindles for rotation of replaceable elements placed in the drive bracket, having a common axis of rotation and, carrying the mixer leash frame, mainly in the form of a square, having diagonally the nodes of fixation and sealing of the mixer flask jacket and hemispherical lids, wherein the ends of the jacket of the mixer flask have conical recesses in which conical protrusions of the hemispherical lids with conical holes at the top of the hemisphere enter the fixing and sealing elements, at least one of which contains a ball valve connected in the fixed position of the said element with the internal volume of the mixer flask.

3. The apparatus for knocking down animal oil from multi-component raw materials on the basis of lactic acid skimmed matrix and alloying additives of vegetable and animal origin, including permanent and replaceable elements, connected kinematically between themselves by mechanisms of rotation and formation of reciprocating motion of knocked down masses of the specified multi-component raw materials, whereby mounting and installation axes of permanent and replaceable elements are at an angle to each other and the axis of rotation of spindles included in the kinematic chain of permanent elements,

4. The apparatus for mixing animal oil containing alloying additives from products of vegetable and animal origin, from initial materials in the form of multicomponent raw materials on the basis of lactic acid matrix with different range of fatness parameters, consisting of a collapsible mixer flask, having at least two centres of fixation, sealing and orientation relative to the mixer leash frame, and one of them has a device for withdrawal of mixing waste and their axes are aligned with the axis of the flask, which is included in a self-contained kinematic chain forming a reciprocating movement of the source material in the mixer flask along its axis, which is at an angle to the axis of rotation of the second self-contained kinematic chain connecting the drive bracket with at least two spindles, of which one is a driving spindle and the other is a driven spindle and a thrust spindle, and furthermore the axes of rotation of both spindles coincide and are horizontal.

5. A process of beating animal oil containing, as a rule, various additives of both vegetable and animal origin, introduced into the initial product before the beginning of the oil beating process, in most cases in liquid form, the initial product being a liquid lactic acid matrix, which, after the introduction of additives, is a multi-component mixture, in which the additives are introduced into the basic lactic acid matrix, comprising successive impact contacting of the said mixture with hemispherical surfaces limiting the inner lactic acid matrix.

Device for pre-mixing and homogenisation of mixture components

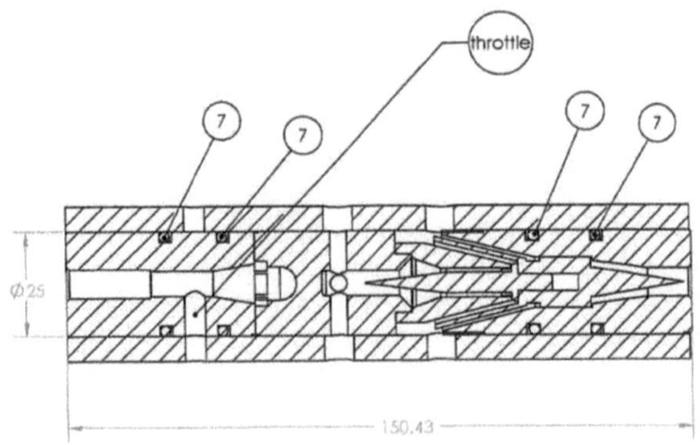
throttle
7
7
7
7
Φ25
150.43

Modern multidisciplinary digital technologies of confectionery production with elements of artificial intelligence and artificial neural networks, psychological nuances and design aspects in full compliance with the requirements and provisions of current standards

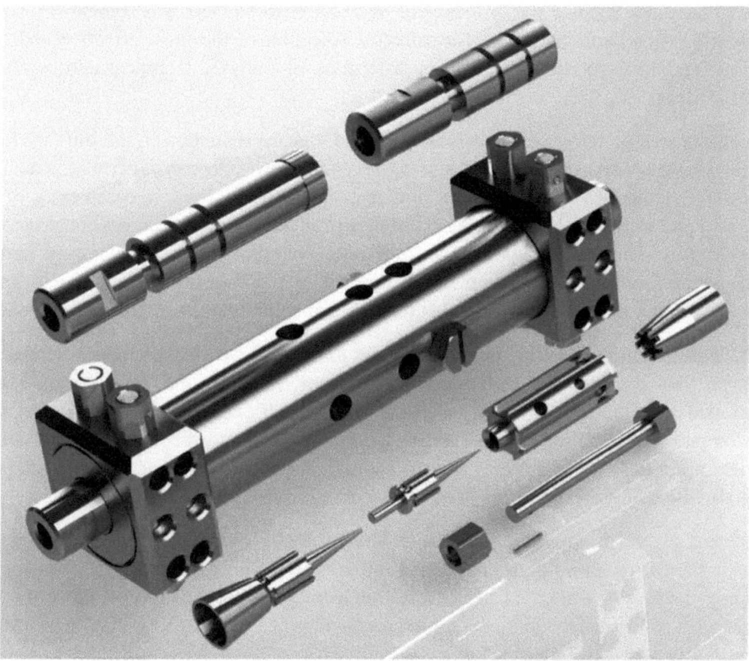

Figure 23 is an example of a modern innovative design with full imitation of the construction materials used and their properties and performance characteristics, including type of machining and type of finish

The model shown in the figure is an exact replica of a device for linear in-line homogenisation of liquid components of confectionery materials and mixtures for subsequent use in the final stages of confectionery production

All inventors know that sometimes there are technical solutions that operate in real conditions, work and solve many problems, which have already pushed the inventor to innovative analysis and initiated his purposeful creative activity, and there are far-fetched obvious technical solutions, which are created in isolation from the real reality and solve absolutely nothing except the realisation of ambitious claims to any (most often useless) idea in the field of engineering and technology; this is especially true in the field of engineering and technology.

In addition, technical solutions arising in any particular local area necessarily directly or indirectly affect the established technical stereotypes and the psychological barriers that have arisen and are constantly arising on their basis, preventing the overcoming of technical and technological contradictions that have arisen on the basis of and in the development of these psychological barriers.

Twenty years ago, the need for inventions of the second group and the equally important need to take into account the influence of psychological barriers was somehow justified by their auxiliary role as a base for selective selection of the most effective and non-obvious technical solutions from the whole mass of initiated technical and combined creative ideas.

The emergence of information technologies and a sharp reduction in the time cycle for the development and transformation of an inventive idea into a product that is actually necessary, demanded by the market and realisable, the increasing complexity of the technical and technological components of new products, causing a proportional increase in the cost of manufacturing prototypes of the invented product and their testing, make it necessary to consider in a completely new way the possibility of creating technical solutions with auxiliary innovative functions, but absolutely unimportant.

Now, if an inventor wants his innovative ideas to be used, he must be more versatile and must possess not only the techniques of foresight, intuition and, to a certain extent, developed imagination, but also be practically a multidisciplinary specialist, at least sensing and (better if) well understanding the commercial and consumer demands of the market, regardless of stereotypes and psychological barriers combined with them, often based on the complete obviousness of the ways to implement synthesised ideas.

There are several basic directions, which have a decisive influence on the fate of new ideas in the current conditions and, taking into account which, may allow to ensure a real and high level of commercial success, or, neglecting which, will forever close the way for the idea to be realised in any commercial form.

The author proposes to review some of these basic trends (naturally, the scope of this publication allows us to do so only in thesis form). :

For modern technological innovations, which are largely formed by active and intensive brainstorming, with the participation of a sufficiently large team or group of specialists interested to a greater or lesser extent in an effective result, which in principle can be considered equivalent to an ideal final result, there are real problems in organising the innovation process, caused by objective reasons, some of which are psychological technological and constructive stereotypes, as well as obvious problems in the organisation of the innovation process.

Emerging and constantly emerging obvious technical - composite - technological solutions are evaluated and considered through the prism of conclusions from the practice of application of the laws of development and construction of industrial technologies, machines and mechanisms, which emerged at least in the last century, at a time when nothing of modern materials, components and component parts in combination with modern industrial electronics and laser technology was not known
In addition, the integration of artificial intelligence elements and artificial neural networks

with the infrastructure of new technical systems makes the processes of design and formation of technical and technological characteristics of innovative products extremely complicated

This situation is further aggravated by the fact that, as mentioned in the first part of this publication, the recognition of a technical solution as an invention is based in the USA and in most other countries on different features, - in the USA on 4 features and in other countries on 3 features.

The mentioned 4 attribute is the subjective factor of technical solution evaluation, which initiates gradual development and entrenchment of psychological stereotype based on division of new technical solutions into obvious and non-obvious ones.

The presence of a clearly expressed subjective factor brings into the process of evaluation of technical solution the comparison with known constructive or technological elements and their combinations familiar from known previous developments

Familiarity of this or that solution and all the nuances of its implementation cannot be an objective factor, as the decision on the obviousness or non-obviousness of a technical solution fundamentally depends on the level of knowledge and professional competence of experts

In modern designs and technological solutions, novelty is not a reflection of a single technical discipline, but of a combination of integrated disciplines including electronics, microelectronics, modern materials science, fibre optics and laser technology, which requires evaluation from all positions of obviousness or non-obviousness, which can only be evaluated by narrow specialists together with specialists in complex integration.

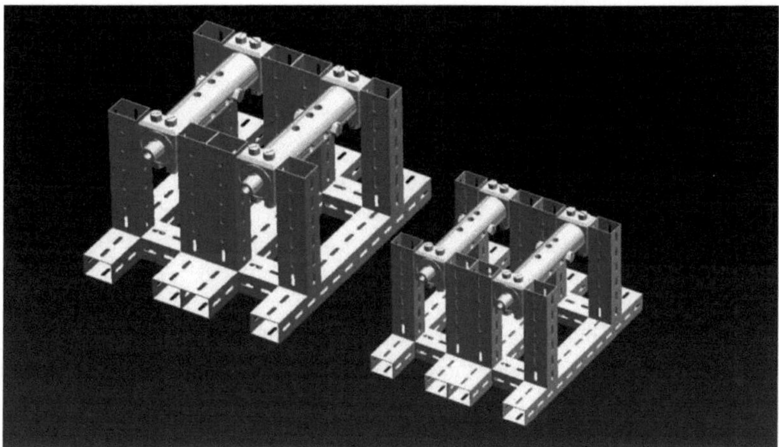

Figure 24, - the figure shows variants of industrial layout of the devices presented in

Figure -1 with three-dimensional analysis of the scale factor, as well as with parallel analysis of the level of non-obviousness of the whole set of solutions associated with this innovative product

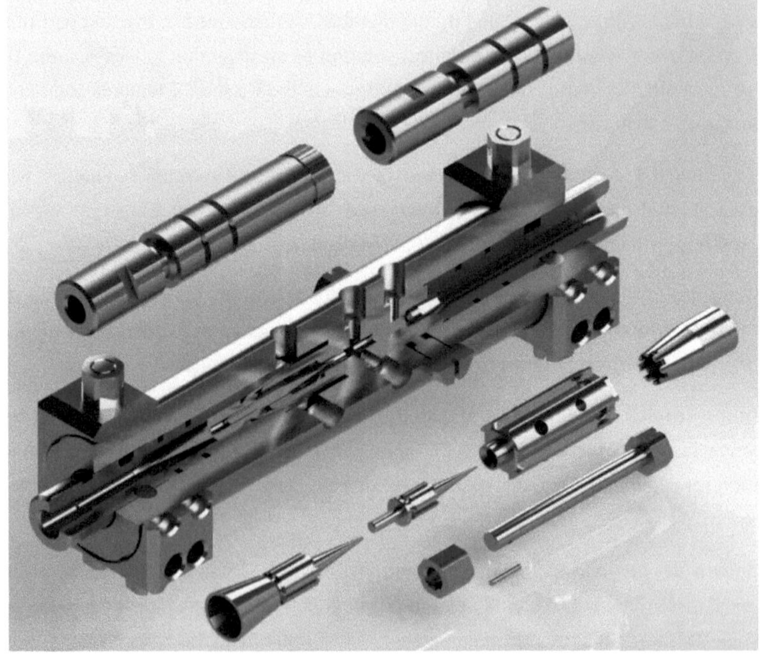

Figure 25 is an example of a modern innovative design using sectional cross sections and with technical and innovative analyses of the level and persuasiveness of qualifying the degree of non-obviousness of all parts of the device and the same degree of non-obviousness of the functional characteristics of the assembly associations and the subsequent commercial advantages of these characteristics
In each discipline, different generic solutions take root over time, gradually evolving into categories of stereotypes familiar to virtually all practitioners
This kind of stereotypes gradually become ingrained in everyday practice and any change in design or technology encounters a certain barrier, again over time turning into a formalised technogenic psychological barrier

In this situation, effective participation in the synthesis of innovative ideas and through the use of brainstorming, bringing these ideas and their innovative combinations to the emergence of a borderline ideal final result may encounter and practically encounter stereotypical psychological barriers, which even in the case of generation of workable technical proposals by part of the group, not obvious to the average specialist in this particular industry, can not be unambiguously perceived and, as a rule, especially in the initial stage, may be rejected.

If we turn to the Theory of Inventive Problem Solving and its methods of achieving an ideal end result, we can see that this theory is not quite adequate in the conditions of modern integrative and complex technical solutions, especially in such complex industries as confectionery production.

However, even without this, the Theory of Inventive Problem Solving had a number of significant flaws, which, obviously, led to the stagnation in its development after the author's death, as well as to significant difficulties in its practical application.

What were these flaws.

In TRIZ, at the level of development of engineering and technology of that time, an attempt was made to formulate the laws of development of technical systems, which were to form the basis of TRIZ and the basis of the general methodology of problem solving.

All these developments remained relevant until the moment of creation and application of processor technology and through them - adaptation with classical ideal results of new conceptual solutions, the effectiveness of which was determined not only by the perfection of constructive and technological solutions and their combinations, but also by the perfection of their adapted software solutions, bringing inventions fundamentally new commercial characteristics

However, most of the regulations and laws formulated 70 years ago are not the same today, due to the widespread adoption of digital technologies.

They should rather be called regularities of technological development, which are far from being complete. This is the reason why there has never been a coherent methodology of problem solving based on the laws of development.

And the formulated laws were mainly used as methodological justifications for the given examples of inventions. The factors of commercial expediency were completely excluded and their influence on the methods of using the constantly modifying laws of development of technical systems for such modernisation, optimisation and modification of the object, which can evolutionarily turn the invention into an unobvious and at a new level of commercial use fully demanded innovative product.

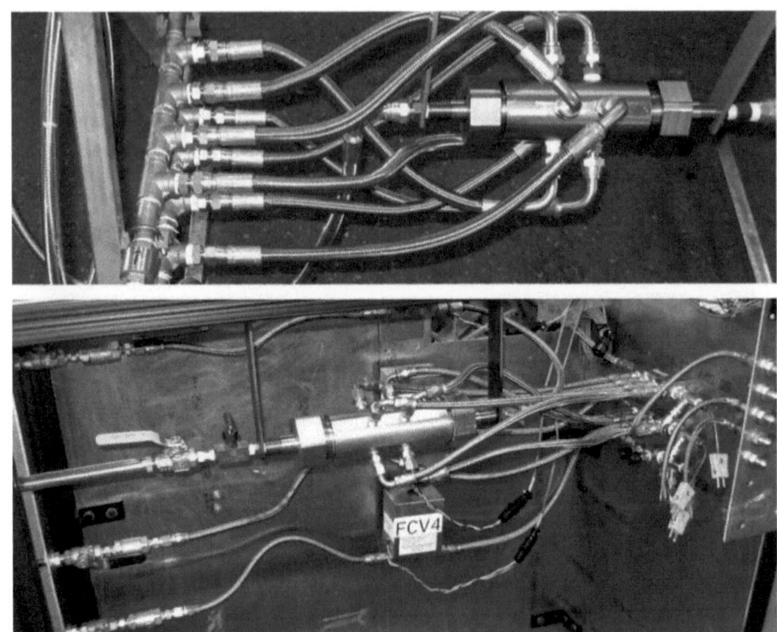

Figures 26 and 27 are examples of combinatorial innovative design based on the principles of overcoming technogenic stereotypes

The recent patent disputes between the world's largest technology companies show and prove that the laws of development of technical systems formulated in TRIZ cannot reflect the whole variety of tasks, functions and features of a modern multifunctional object that is not obvious in its structural characteristics, and taking into account all the new factors that have emerged and constantly continue to emerge, characterising an innovative object, it is necessary to redefine these laws by linking them to the laws of development of commercial structures and to the laws of development of innovation.

1. The dialectical approach (analysis of contradictions) embedded in the main tool of problem solving, which was ARIZ, was supplemented and largely changed by the introduction of new concepts (technical and physical contradiction).
2. These new notions changed the essence of the dialectical contradiction formulated in dialectical logic, which led to difficulties in identifying the contradiction when trying to solve real inventive problems with the help of ARIZ and did not qualify the level of non-obviousness of the technical solution, especially when adapting with elements of artificial intelligence and artificial neural networks.

In view of all the digital technologies and software applications implemented with basic classical solutions, we should focus on this separately - what can be considered a real inventive task, how to harmonise the psychological combination of evaluations from the inventor, manufacturer and consumer?

How can a correct or incorrect formulation of an inventive problem affect the commercialisation of the resulting invention ?

Is it even possible to reliably protect the resulting complex technical solution from unauthorised copying ?

The search for answers to all these and many other questions is now becoming a major part of the dialectic of creating a strategy for patenting and licensing inventions, certainly in conjunction with qualifying the degree of non-obviousness of the invention.

Improvement of ARIZ (creation of new modifications of the algorithm from ARIZ-77 to ARIZ-85B) went not along the path of eliminating inaccuracies in the procedures of contradiction detection, but along the path of complicating the algorithm.

As a result, the last official modification of the ARIZ-85B algorithm turned into an extremely cumbersome and of little practical use.

3. TRIZ has never found clear mechanisms of transition from a formulated contradiction to its practical resolution. This created serious difficulties in solving real-world problems with the help of TRIZ.
4. TRIZ declared the rejection of the methodology of activating the search of variants, but the main part of the so-called TRIZ tools represented exactly such methods.
5. In TRIZ, such analysis was presented as a scientific approach based on the analysis of regularities of structural development of technical objects. However, the assumption that non-existent physical fields are used in the analysis, as well as the possibility of ambiguous interpretation of designs and rules of their transformation, rather allow us to refer this analysis to the methods of activating the selection of variants, but not to scientific analysis.
6. The closest to the idea of formalising the procedure of inventive problem solving was the creation in TRIZ of tables and techniques for resolving technical contradictions. This approach was based on the statistical analysis of the then existing invention descriptions. However, despite the existing prospects, it was not further developed in TRIZ, and due to a number of shortcomings and obsolescence of statistical conclusions, it lost its relevance for practical use.
7. There is a widespread illusion that TRIZ can be implemented in real production. In its essence TRIZ is an individual method of problem solving, the application of which is a personal choice for a person, i.e. there is a serious psychological aspect to personal choice. For this reason it is impossible to make TRIZ a part of any production process. At best, an enterprise can organise TRIZ training for its employees in order to enhance their creative capabilities (which is done today in the

USA).

During the period of its active development (the 1980s), these shortcomings and errors were successfully compensated by the enthusiasm of TRIZ adherents. Nevertheless, the existing flaws of TRIZ and the departure from TRIZ as a result of the production crisis of its main developers, who were able to see these flaws, led to a stagnation in the development of the theory. In the opinion of the author of this publication, this is the main reason why nothing new worthy of serious attention has appeared in TRIZ in the last decade.

List of used literature, patent and licence information

APPENDIX 1-1

United States Patent Application	**20210104744**
Kind Code	**A1**
OGUNI; Teppei ; et al.	**April 8, 2021**

SECONDARY BATTERY AND MANUFACTURING METHOD THEREOF

Abstract

Provided is a layer for preventing a short circuit between a positive electrode and a negative electrode in a solid ***battery*** using a layer containing a solid electrolyte. As the solid electrolyte between the positive electrode and the negative electrode, a layer containing a graphene compound is used. Lithium ions can pass through the layer containing the graphene compound. Lithium ions are added in advance in the layer containing the graphene compound. Specifically, a modifier is used, and a graphene compound chemically modified with a functional group such as ether and ester with an increased interlayer distance is used.

APPENDIX 1-2

United States Patent Application	**20210104719**
Kind Code	**A1**
OIKAWA; Makiko	**April 8, 2021**

BATTERY ELECTRODE AND METHOD FOR MANUFACTURING THE SAME

Abstract

There is provided a method for manufacturing a ***battery*** electrode. The method includes: forming a precursor of the ***battery electrode*** including a double-sided coating area in which both sides of a current collector are coated with an electrode material layer and a single-sided coating area adjacent to the double-sided coating area; subjecting the current collector located at a boundary portion between the double-sided coating area and the single-sided coating area to a heat treatment locally; and pressurizing the precursor of the ***battery*** electrode. The single-sided coating area includes a main side of the current collector that is coated with the electrode material layer.

APPENDIX 1-3

United States Patent Application **20210100579**
Kind Code **A1**
Shelton, IV; Frederick E. ; et al. **April 8, 2021**

MODULAR BATTERY POWERED HANDHELD SURGICAL INSTRUMENTS AND METHODS THEREFOR

Abstract

Disclosed is a method of controlling a modular ***battery*** powered handheld surgical instrument. The surgical instrument including a ***battery, a*** user input sensor, a controller, a radio frequency (RF) drive circuit, an ultrasonic transducer, an ultrasonic transducer drive circuit, and an end effector. The end effector including an electrode electrically coupled to the RF drive circuit, an ultrasonic blade acoustically coupled to the ultrasonic transducer, and a sensor to measure tissue parameters. The method includes applying an RF current drive signal to the electrode by the RF drive circuit; applying an ultrasonic drive signal to the ultrasonic transducer by the ultrasonic transducer drive circuit to acoustically excite the ultrasonic blade; controlling intensity, wave shape, and/or frequency of the RF current drive signal and the ultrasonic drive signal on a sensed measure of a tissue or user parameter.

APPENDIX 1-4

United States Patent Application **20210091416**
Kind Code **A1**
SEKI; Hayato ; et al. **March 25, 2021**

SECONDARY BATTERY, BATTERY PACK, VEHICLE, AND STATIONARY POWER SUPPLY

Abstract

A secondary ***battery*** includes a positive electrode, a first aqueous electrolyte held on the positive electrode, a negative electrode, a second aqueous electrolyte held on the negative electrode, and a separator interposed between the positive electrode and the negative electrode. A difference between an osmotic pressure (N/m.sup.2) of the first aqueous electrolyte and an osmotic pressure (N/m.sup.2) of the second aqueous electrolyte is 90% or less (including 0%) of the higher one of the osmotic pressure of the first aqueous electrolyte and the osmotic pressure of the second aqueous electrolyte.

APPENDIX 1-5

United States Patent Application	**20210091402**
Kind Code	**A1**
Londarenko; Yuriy Y.	**March 25, 2021**

MULTI-LAYER BATTERY CONFIGURATIONS

Abstract

Rechargeable ***battery*** cells according to embodiments of the present technology may include a ***housing*** including a first conductive segment operable at anode potential, and a second conductive segment operable at cathode potential. The ***housing*** may include a gasket positioned between the first conductive segment and the second conductive segment and configured to hermetically seal the ***housing***. The ***battery*** cells may also include an electrode stack. The electrode stack may include a cathode current collector having a cathode active material extending across a first surface of the cathode current collector. The cathode current collector may be characterised by at least two pleats. The electrode stack may also include an anode current collector having an anode active material extending across a first surface of the anode current collector. The anode current collector may be characterised by at least two pleats.

APPENDIX 1-6

United States Patent Application	**20210091363**
Kind Code	**A1**
Lane; Robert Clinton	**March 25, 2021.**

High Voltage Battery Module Parallel Cell Fusing System

Abstract

A fusing system for a brick of lithium ion ***battery in a battery*** module is provided where the fusing system has a combination of low-voltage fuses and a high-voltage fuse. The low-voltage fuse can have one or more fusing elements in a springy spiral configuration or a straight configuration with the fuse element encapsulated.

APPENDIX 1-7

United States Patent Application	**20210091360**
Kind Code	**A1**
Balaram; Haran ; et al.	**March 25, 2021**

BATTERY PACK WITH ATTACHED SYSTEM MODULE

Abstract

Battery systems according to embodiments of the present technology may include a ***battery***. The ***battery*** may include a first electrode terminal and a second electrode terminal accessible along a first surface of the ***battery***. The systems may include a module electrically coupled with the ***battery***. The module may include a circuit board characterised by a first surface and a second surface opposite the first surface. The module may include a mould extending from the first surface of the circuit board towards the ***battery***. The module may include a first conductive tab electrically coupling the module with the first electrode terminal. The module may include a second conductive tab electrically coupling the module with the second electrode terminal. The second conductive tab may extend across the mould substantially parallel to the first surface of the circuit board.

APPENDIX 1-8

United States Patent Application **20210075063**
Kind **CodeA1**
Dou; Shushi ; et al. **March 11, 2021**

LITHIUM-ION BATTERY AND APPARATUS

Abstract

This application provides a lithium-ion ***battery*** and an apparatus. The lithium-ion ***battery*** includes an electrode assembly and an electrolyte. The electrode assembly includes a positive electrode plate, a negative electrode plate, and a separator. A positive active material of the positive electrode plate includes Li.sub.x1Co.sub.y1M.sub.1-y1O.sub.2-z1Q.sub.z1, where 0.5.ltoreq.x1.ltoreq.1.2, 0.8.ltoreq.y1.ltoreq.1.0, 0.ltoreq.z1.ltoreq.0.1, M is selected from one or more of Al, Ti, Zr, Y, and Mg, and Q is selected from one or more of F, Cl, and S. The electrolyte contains an additive A, an additive B, and an additive C. The additive A is a polynitrile six-membered nitrogen-heterocyclic compound with a relatively low oxidation potential. The additive B is a silyl phosphite compound or a silyl phosphate compound or a mixture thereof. The additive C is a halogen substituted cyclic carbonate compound.

APPENDIX 1-9

United States Patent Application **20210075060**
Kind Code **A1**
NAKAYAMA; Tetsuri **March 11, 2021**

NON-AQUEOUS ELECTROLYTE SECONDARY BATTERY

Abstract

A non-aqueous electrolyte secondary ***battery*** disclosed herein includes a positive electrode, a negative electrode, and a non-aqueous electrolyte. The positive electrode includes a positive electrode current collector and a positive electrode active material layer provided on the positive electrode current collector. The non-aqueous electrolyte contains lithium fluorosulfonate. The positive electrode active material layer contains a positive electrode active material. The positive electrode active material layer contains hydrated alumina at least in a surface layer portion.

APPENDIX 1-10

United States Patent Application	**20210075015**
Kind Code	**A1**
LEE; Jungmin ; et al.	**March 11, 2021**

SECONDARY LITHIUM BATTERY ANODE AND SECONDARY LITHIUM BATTERY INCLUDING THE SAME

Abstract

Disclosed are a secondary lithium ***battery*** anode and a secondary lithium ***battery*** including the same, the secondary lithium ***battery*** anode comprising a current collector and an anode active material layer located on at least one surface of the current collector, wherein the anode active material layer includes an anode active material layer which has sphericity of 0.83 to 0.91, and a binder which has an average particle diameter (D50) of 180 nm to 450 nm.

ANNEX 1-11

United States Patent Application	**20210075008**
Kind Code	**A1**
Park; Benjamin Yong ; et al.	**March 11, 2021**

Prelithiated And Methods For Prelithiating An Energy Storage Device

Abstract

The present disclosure relates to prelithiated Si electrodes, methods of prelithiating Si electrodes, and use of prelithiated electrodes in electrochemical devices are described. There are several characteristics of electrode prelithiaiton that enable the superior ***battery*** performance. First, a prelithiated silicon anode is already in its expanded state during SEI formation, and therefore less of the SEI layer breaks down and reforms during cycling. Second, the prelithiated anode has a lower anode potential, which may also help the cycle performance of an electrochemical device.

ANNEX 1-12

United States Patent Application **20210066712**
Kind Code **A1**
KIM; Young-Ki ; et al. **March 4, 2021.**

POSITIVE ELECTRODE ACTIVE MATERIAL FOR RECHARGEABLE LITHIUM BATTERY AND RECHARGEABLE LITHIUM BATTERY INCLUDING SAME

Abstract

The present invention relates to a positive electrode active material for a rechargeable lithium ***battery*** and a rechargeable lithium ***battery*** including same, wherein the positive electrode active material comprises a core and a surface layer formed on the surface of the core, the core comprising a first crystalline structure, the surface layer comprising a first crystalline structure and a second crystalline structure different from the first crystalline structure, the first crystalline structure being present more than the second crystalline structure in the surface layer.

ANNEX 1-13

United States Patent Application **20210066683**
Kind Code **A1**
Lane; Robert Clinton **March 4, 2021.**

Method to Prevent or Minimize Thermal Runaway Events in Lithium Ion Batteries

Abstract

A method to prevent or minimise an occurrence of a thermal runaway event in a ***battery*** module of an electric vehicle. The method places a gas barrier between a venting space and a wall of each ***battery*** cell so that escaped gas from one ***battery*** cell does not impinge onto another ***battery*** cell.

ANNEX 1-14

United States Patent Application	**202100577777**
Kind Code	**A1**
SUGIYO; Takeshi ; et al.	**February 25, 2021**

ALL-SOLID-STATE BATTERY, METHOD FOR MANUFACTURING THE SAME, AND PROCESSING DEVICE

Abstract

The present invention prevents edge collapse of an electrode layer of a laminated body included in an all-solid-state ***battery***. A method of producing an all-solid-state ***battery*** includes: a laminated body forming step of forming a laminated body (310) including (i) a positive electrode layer (302), (ii) a negative electrode layer (304) having a polarity opposite of a polarity of the positive electrode layer (302); and (iii) a solid-electrolyte layer (303) disposed between the positive electrode layer (302) and the negative electrode layer (304); and a cutoff step of cutting off an outer peripheral edge of the laminated body (310) so as to form a laminated body containing a powder material.

ANNEX 1-15

United States Patent Application	**20210057755**
Kind Code	**A1**
Brewer; John C. ; et al.	**February 25, 2021**

ANODES FOR LITHIUM-BASED ENERGY STORAGE DEVICES

Abstract

An anode for a lithium-based energy storage device such as a lithium-ion ***battery*** is disclosed. The anode includes a current collector having an electrically conductive layer and a surface layer overlaying the electrically conductive layer. A lithium storage layer is overlaying the surface layer and the surface layer includes a metal chalcogenide having at least one of sulfur or selenium. The metal chalcogenide may include a metal sulfide, a metal polysulfide, a metal selenide, a metal polyselenide, or a combination thereof. The metal chalcogenide may include a copper sulfide or a copper polysulfide. The lithium storage may include a total content of silicon, germanium, or a combination thereof of at least 40 atomic %. The lithium storage layer may be a continuous porous lithium storage layer having an average density from about 1.1 g/cm.sup.3 to about 2.25 g/cm.sup.3 and comprises at least 85 atomic % amorphous silicon.

ANNEX 1-16

United States Patent Application **20210057721**
Kind Code **A1**
KAWASAKI; Daisuke ; et al. **February 25, 2021**

LITHIUM ION SECONDARY BATTERY

Abstract

Provided is a lithium ion secondary ***battery*** having high energy density and excellent cycle characteristics, and hardly causing burning. The present invention relates to a lithium ion secondary ***battery*** comprising an electrode mixture layer comprising an electrode active material comprising Si alloy having a median diameter of 1.2 .mu.m or less and 12% by weight or more and 50% by weight or less of an electrode binder; and an electrolyte solution comprising 60% by volume or more and 99% by volume or less of a phosphoric acid ester compound, 0% by volume or more and 30% by volume or less of a fluorinated ether compound, and 1% by volume or more and 35% by volume or less of a fluorinated carbonate compound, wherein the total amount of the phosphoric acid ester compound and the fluorinated ether compound is 65% by volume or more.

ANNEX 1-17

United States Patent Application **20210057690**
Kind Code **A1**
Fukutome; Kazuaki ; et al. **February 25, 2021**

BATTERY PACK AND METHOD FOR PRODUCING THE SAME

Abstract

A ***battery*** pack includes a plurality of secondary ***battery*** cells, a ***battery*** holder, a circuit substrate, and an outer case, in which the ***battery*** holder is divided into a plurality of divided holders, each of the divided holders forms a fitting structure for fitting the divided holders to each other at an interface for joining the divided holders together, the ***battery*** holder forms a substrate holding area that holds a circuit substrate with the circuit substrate surrounded by side walls in a state where the divided holders are coupled to each other by the fitting structure, a joint interface where the divided holders are fitted to each other by the fitting structure of the divided holders is exposed in the substrate holding area, and a surface of the circuit substrate is covered with a potting resin in the substrate holding area.

ANNEX 1-18

United States Patent Application	**20210057687**
Kind Code	**A1**
MATSUO; Tatsumi ; et al.	**February 25, 2021**

BATTERY DEVICE AND MANUFACTURING METHOD

Abstract

A ***battery*** device includes a ***battery*** cell; an exterior member that accommodates the ***battery cell***; and one or more foamed resin fixing members disposed between the ***battery*** cell and the exterior member. The foamed resin fixing members are formed of a foamed resin having self-adhesiveness.

ANNEX 1-19

United States Patent Application	**20210053689**
Kind Code	**A1**
Lynn; Robert ; et al.	**February 25, 2021**

VEHICLE CABIN THERMAL MANAGEMENT SYSTEM AND METHOD

Abstract

The system can include an on-board thermal management subsystem. The system 100 can optionally include an off-board (extravehicular) infrastructure subsystem. The onboard thermal management subsystem can include: a ***battery*** pack, one or more fluid loops, and an air manifold. The system 100 can additionally or alternatively include any other suitable components.

ANNEX 1-20

United States Patent Application	**20210052313**
Kind Code	**A1**
Shelton, IV; Frederick E. ; et al.	**February 25, 2021**

MODULAR BATTERY POWERED HANDHELD SURGICAL INSTRUMENT WITH SELECTIVE APPLICATION OF ENERGY BASED ON TISSUE CHARACTERISATION

Abstract

A surgical instrument comprises a shaft assembly comprising a shaft and an end effector coupled to a distal end of the shaft; a handle assembly coupled to a proximal end of the shaft; a ***battery assembly*** coupled to the handle assembly; a radio frequency (RF) energy output powered by the ***battery*** assembly and configured to apply RF energy to a tissue; an ultrasonic energy output powered by the ***battery*** assembly and configured to apply ultrasonic energy to the tissue; and a controller configured to, based at least in part on a measured tissue characteristic, start application of RF energy by the RF energy output or application of ultrasonic energy by the ultrasonic energy output at a first time.

ANNEX 1-21

United States Patent Application **20210050591**
Kind Code **A1**
Brewer; John C. ; et al. **February 18, 2021**

ANODES FOR LITHIUM-BASED ENERGY STORAGE DEVICES, AND METHODS FOR MAKING THE SAME

Abstract

A method of making a prelithiated anode for use in a lithium-ion ***battery*** includes providing a current collector having an electrically conductive layer and a metal oxide layer overlaying the electrically conductive layer. The metal oxide layer has an average thickness of at least 0.01 .mu.m. A continuous porous lithium storage layer is deposited onto the metal oxide layer by a CVD process. Lithium is incorporated into the continuous porous lithium storage layer to form a lithiated storage layer prior to a first electrochemical cycle when the anode is assembled into the ***battery***. The anode may be incorporated into a lithium ion ***battery*** along with a cathode. The cathode may include sulfur or selenium and the anode may be prelithiated.

ANNEX 1-22

United States Patent Application **20210043941**
Kind Code **A1**
HORIUCHI; Hiroshi ; et al. **February 11, 2021.**

BATTERY

Abstract

A ***battery*** includes a positive electrode that includes a positive electrode current collector and a positive electrode active material layer provided on the positive electrode current collector and has a positive electrode current collector exposed portion at which the positive electrode current collector is exposed; a negative electrode that includes a negative electrode current collector and a negative electrode active material layer provided on the negative electrode current collector and has a negative electrode current collector exposed portion at which the negative electrode current collector is exposed; a separator provided between the positive electrode and the negative electrode; and an intermediate layer that is provided between the separator and at least one of the positive and negative electrodes and includes at least one of a fluororesin and a grain.

ANNEX 1-23

United States Patent Application	**20210043931**
Kind Code	**A1**
KIM; Do-Yu	**February 11, 2021**

POSITIVE ACTIVE MATERIAL PRECURSOR FOR RECHARGEABLE LITHIUM BATTERY, POSITIVE ACTIVE MATERIAL FOR RECHARGEABLE LITHIUM BATTERY, METHOD OF PREPARING THE POSITIVE ACTIVE MATERIAL, AND RECHARGEABLE LITHIUM BATTERY INCLUDING THE POSITIVE ACTIVE MATERIAL

Abstract

An embodiment provides a positive active material precursor for a rechargeable lithium ***battery*** including: a nickel-based ***composite*** precursor including a secondary particle comprising a plurality of primary particles that are aggregated together, the nickel-based ***composite precursor*** having a central portion and a surface portion, and the central portion of the nickel-based ***composite precursor*** including a phosphate.

ANNEX 1-24

United States Patent Application	**20210036368**
Kind Code	**A1**
JIANG; Yao ; et al.	**February 4, 2021**

LITHIUM-ION BATTERY AND APPARATUS

Abstract

This application provides a lithium-ion ***battery*** and an apparatus. The lithium-ion ***battery*** includes an electrode assembly and an electrolyte. The electrode assembly includes a positive electrode plate, a negative electrode plate, and a separator. A positive active material of the positive electrode plate includes Li.sub.x1CO.sub.y1M.sub.1-y1O.sub.2-z1Q.sub.z1, where 0.5.ltoreq.x1.ltoreq.1.2, 0.8.ltoreq.y1.ltoreq.1.0, 0.ltoreq.z1.ltoreq.0.1, M is selected from one or more of Al, Ti, Zr, Y, and Mg, and Q is selected from one or more of F, Cl, and S. The electrolyte contains an additive A, an additive B, and an additive C. The additive A is a polynitrile six-membered nitrogen-heterocyclic compound with a relatively low oxidation potential. The additive B is an anhydride compound. The additive C is a halogen substituted cyclic carbonate compound.

Printed by Books on Demand GmbH, Norderstedt / Germany